Saving the Forests: What Will It Take?

ALAN THEIN DURNING

Nancy Chege, *Staff Researcher*

Carole Douglis, *Editor*

WORLDWATCH PAPER 117
December 1993

THE WORLDWATCH INSTITUTE is an independent, nonprofit environmental research organization based in Washington, D.C. Its mission is to foster a sustainable society—in which human needs are met in ways that do not threaten the health of the natural environment or future generations. To this end, the Institute conducts interdisciplinary research on emerging global issues, the results of which are published and disseminated to decisionmakers and the media.

FINANCIAL SUPPORT is provided by the Geraldine R. Dodge Foundation, W. Alton Jones Foundation, John D. and Catherine T. MacArthur Foundation, Andrew W. Mellon Foundation, Edward John Noble Foundation, Pew Charitable Trusts, Lynn R. and Karl E. Prickett Fund, Rockefeller Brothers Fund, Surdna Foundation, Turner Foundation, Frank Weeden Foundation, Joyce Mertz-Gilmore Foundation, Energy Foundation, and Wallace Genetic Foundation.

PUBLICATIONS of the Institute include the annual *State of the World*, which is now published in 27 languages; *Vital Signs*, an annual compendium of the global trends—environmental, economic, and social—that are shaping our future; the *Environmental Alert* book series; and *World Watch* magazine, as well as the *Worldwatch Papers*. For more information on Worldwatch publications, write: Worldwatch Institute, 1776 Massachusetts Ave., N.W., Washington, DC 20036; or FAX (202) 296-7635.

THE WORLDWATCH PAPERS provide in-depth, quantitative and qualitative analysis of the major issues affecting prospects for a sustainable society. The Papers are authored by members of the Worldwatch Institute research staff and reviewed by experts in the field. Published in five languages, they have been used as a concise and authoritative reference by governments, nongovernmental organizations and educational institutions worldwide. For a partial list of available Papers, see page 53.

Table of Contents

Tables and Figures

Sections of this paper may be reproduced in magazines and newspapers with written permission from the Worldwatch Institute. For information, call Director of Communications, at (202) 452-1999 or FAX (202) 296-7365.

The views expressed are those of the author, and do not necessarily represent those of the Worldwatch Institute and its directors, officers, or staff, or of funding organizations.

Introduction

Imagine a time-lapse film of the Earth taken from space. Play back the last 10,000 years sped up so that a millennium passes by every minute. For more than seven of the ten minutes, the screen displays what looks like a still photograph: the blue planet Earth, its lands swathed in a mantle of trees. Forests cover 34 percent of the land. Aside from the occasional flash of a wildfire, none of the natural changes in the forest coat are perceptible. The Agricultural Revolution that transforms human existence in the film's first minute is invisible.[1]

After seven and a half minutes, the lands around Athens and the tiny islands of the Aegean Sea lose their forest. This is the flowering of classical Greece. Little else changes. At nine minutes—1,000 years ago—the mantle grows threadbare in scattered parts of Europe, Central America, China, and India. Then 12 seconds from the end, two centuries ago, the thinning spreads, leaving parts of Europe and China bare. Six seconds from the end, one century ago, eastern North America is deforested. This is the Industrial Revolution. Little else appears to have changed. Forests cover 32 percent of the land.[2]

In the last three seconds—after 1950—the change accelerates explosively. Vast tracts of forest vanish from Japan, the Philippines, and the mainland of Southeast Asia, from most of Central America and the horn of Africa, from western North America and eastern South America, from the Indian subcontinent and sub-Saharan Africa. Fires rage in the Amazon basin where they never did before, set by ranchers and peasants. Central Europe's forests die, poisoned by the air and the rain. Southeast Asia resembles a dog with mange. Malaysian Borneo appears shaved. In the final fractions of a second, the clearing spreads to Siberia and the Canadian north. Forests disappear so suddenly from so many places that it looks like a plague of locusts has descended on the planet.

The film freezes on the last frame. Trees cover 26 percent of the land. Three-fourths of the original forest area still bears some tree cover. But just 12 percent of the earth's surface—one-third of the initial total—consists of intact forest ecosystems. The rest holds biologically impoverished stands of commercial timber and fragmented regrowth. This is the present: a globe profoundly altered by the workings—or failings—of the human economy.[3]

On the ground, this wave of deforestation has created a paradox: rising economic prosperity amid declining ecological health. At some point, probably long passed, the ecological costs of deforestation exceeded its economic benefits. But deforestation has continued.

In nature's economy, forests play pivotal roles. They provide habitat for perhaps half of the 10 to 80 million forms of life on Earth.[4] (The range is wide because biologists have yet to catalog most of the planet's plant and animal species.) Forests hold much of the genetic information accumulated during 4 billion years of evolution. They buffer the global climate against greenhouse warming, and moderate local climes. They prevent floods and droughts, keep pests in check, filter the air, secure the soil and keep it out of waterways, and nurture fisheries in rivers and lakes.

In the money economy, however, forests show up only as sources of lumber and fuel, or as thickets to be cleared for new farms and suburbs. Predictably, if tragically, this misalignment between the natural economy and the human economy has resulted in losses for both.

Natural scientists have detailed the costs to ecosystems: extinctions of thousands of species, worsened droughts and floods, release of heat-trapping carbon dioxide, widened extremes in local temperatures, introduction of new agricultural pests, erosion of topsoil, clogging of rivers and hydroelectric reservoirs, and loss of fisheries.

Social scientists have documented the human costs. Logging towns in the Pacific Northwest of the United States are showing the signs of decline long evident in inner cities: alcoholism and drug abuse, domestic violence and broken families, homelessness and emigration. Indigenous cultures in tropical forests are wracked by violence, disease, and alienation. Villages in the woods of Africa and Asia suffer falling crop yields and shrinking game supplies.

Each of these litanies has been thoroughly studied and widely reported. What is not known, however, is what to do. Already, a string of well-meaning initiatives have failed. The Tropical Forestry Action Plan, the International Tropical Timber Agreement, the International Tropical Timber Organization, and the United Nations' Statement of Forest Principles were each launched with fanfare and high hopes. They have each proved disappointing, if not fruitless. Deforestation rates in much of the world have continued to rise.[5]

Past efforts foundered because they set their sights on the proximate rather than ultimate causes of deforestation.

Halting deforestation requires no less than restructuring three features of the economy: property rights to forests, pricing of forest products, and political power over the disposition of forests.

Previous policies sought to halt those who fell the trees—the commercial cultivators, ranchers, real estate speculators, developers, loggers, miners, and slash-and-burn farmers. But these are merely the leading edge of a larger force. They are the teeth of the saw; the saw is a money economy blind to its ecological roots.

What would it really take to save the forests? Halting deforestation requires no less than restructuring three features of the economy: property rights to forests, pricing of forest products, and political power over the disposition of forests.

- *Property rights* laws define the terms of ownership and use of forests. But in much of the world, they leave the day-to-day users and managers of the trees with little incentive to care about the ecological health of the forests.

- *Prices* of wood and other forest products do not tell the ecological truth. Soil erosion, decimated fisheries, and other environmental costs of production do not show up in market prices, leaving timber underpriced and forests undervalued.
- Political *power* over forests is concentrated in the hands of the few who profit from deforestation, not in the hands of the many who depend on the forests' health.

Correcting these structural flaws would realign the money economy with nature's economy. Such changes would also dramatically alter the way we use forests, no less for Northern professionals who read a pound of newsprint each morning than for Southern settlers who feed themselves from rainforest plots. But neglecting property rights, price, and power would condemn the forests, and their half-billion human inhabitants, to continuing decline.

Rights to the Forests

The first prerequisite for a sustainable forest economy is a property rights system that allies the interests of forest people with the health of forest ecosystems. This need not mean private ownership. Effective property rights systems, or "tenure regimes," range from private ownership to collective management by communities to state control. They also range from the inclusive—covering land, trees, wildlife, and mineral deposits in perpetuity—to the limited, covering particular activities in certain places and times. In some African villages, for example, households are allotted use of specific farm plots, hunting trails, fruit trees, and even wild bee hives; in return, they must be good stewards and share the bounty as social convention dictates. No single property rights system is always best. What matters is that governments match tenure laws with the social context. At present, there are great disparities, as the following case illustrates.

Four decades ago, the Aravali hills that surround the city of

Udaipur in the Indian state of Rajasthan were a vast forested zone of small villages. Tigers roamed and water flowed in streams year-round. Today, the hills are barren: Their tigers are gone, and their streams flood and parch with the seasons. The villagers' farms are less productive, and thousands of their able-bodied sons and daughters drift from city to city as migrant laborers. "Our region is a case study in ecological decline," says Jagat Mehta, president of Seva Mandir, a voluntary development organization that has worked in the area for 30 years.[6]

Neglecting property rights, price, and power would condemn the forests, and their half-billion human inhabitants, to continuing decline.

In the early eighties, Seva Mandir began helping villagers plant trees. As in all too many such initiatives, few of the seedlings survived. The immediate problem was goats, which ate the seedlings before they could take root. But there was a deeper problem as well. Most of the villages were surrounded by land classified as "forest;" consequently that land fell under the exclusive jurisdiction of the state forest department. Indeed, if a tree were ever to grow there, regardless of who planted or cared for it, it would belong to the forest department. No wonder the seedlings ended up as goat fodder.

After the initial attempts to plant trees, Seva Mandir petitioned the government to give villagers a stake in managing the denuded forestland by creating a "joint forest management" agreement. Under joint forest management, pioneered in eastern India, governments continue to own forestlands, but villagers receive long-term rights to use them. At first, Seva Mandir's pleas fell on the deaf ears of the bureaucracy: The state government had no procedure for granting such arrangements. Eventually, however, Seva Mandir's persistence paid off, and the group gained several experimental joint forest management agreements. As a result, the government allowed villagers to cut trees and harvest other products grown on forestland.

By 1991, the results were drawn in patches of dark green on

the hills around Aravali villages such as Dolpura. Trees were growing again—shoulder-high and surrounded with thick grass. Barriers of dried thor, a type of cactus, formed the first defense against goats. Watchful villagers scanned the perimeter for breaches in the line, vigilant to protect their trees. This "social fencing" was the only kind that worked against the goats as well as their owners, who might otherwise try to slip in for a little fuelwood or fodder. The return of forests was received with such appreciation that the people of Dolpura declared the first restored hill a sacred grove, forbidding any felling there.

Success in places like Dolpura has been striking. So striking, in fact, that forest departments in many Indian states have had little choice but to share control of forests with more villages. By late 1993, 13 Indian states had issued official edicts in support of joint management. As many as 10,000 villages were sharing management responsibilities in an area of perhaps 1.5 million hectares. The concept has been gaining popularity elsewhere in Asia and Africa: In the Philippines, for example, residents manage 171,000 hectares of upland and mangrove forest. And villagers have been negotiating informal agreements with local foresters across Southeast Asia.[7]

The spread of joint management programs is full of historical irony. It is a tacit admission that forest policies in the tropics are ill-founded. In India, as elsewhere, colonial powers nationalized forests to expedite the extraction of timber. Deeming forest dwellers incompetent to care for the lands they had inhabited for centuries, authorities established forest departments on the European model: guards to keep commoners out of royal hunting reserves. In the space of just 150 years, 80 percent of the world's tropical forests passed from the hands of local communities—the Dolpuras of the world—to the hands of government foresters and other officials.[8]

Needless to say, the transfer never gained the assent of most forest dwellers. Linka Ansulang, a tribal woman from the southern Philippines, speaks for many when she says, "The government calls us squatters, but the government is the real squatter."[9]

After independence, Third World leaders continued the centralized model of forest management, believing, as had their predecessors, that modern nations could not entrust illiterate villagers with the fate of forests. Over time, a theory of resource use—the "tragedy of the commons"—emerged that provided a convenient justification for the policy of state control. A tragedy of the commons arises when lack of clear property rights leads individuals to overexploit shared resources or risk losing out while others do so. In such a situation, so the reasoning goes, only a central authority can reconcile individual and social interests. And "central authority" is usually interpreted to mean the state.

What the theory overlooked was that under village control, forests were not unregulated commons. In Dolpura, as in most of the world's thousands of forest communities, villagers organized elaborate systems to manage use of shared forests. In Borneo, for example, generations of the Galik people carefully conserved ironwood trees, allotting rights along kinship lines. Ironwood yields the world's heaviest lumber, prized for its durability and resistance to insects. But the state considers the lands of the Galik national property. So when state-sanctioned loggers entered the area in the early nineties, the Galik saw they would soon lose the ironwood to the loggers—and raced to sell it themselves.[10]

The spread of joint management programs is a tacit admission that forest policies in the tropics are ill-founded.

In Borneo, as elsewhere, nationalizing the forests sabotaged traditional management, creating the free-for-all it purported to avert. The government made it a crime to impose local penalties for overuse of forests. Yet national penalties for the same transgressions, even if written in law, proved unenforceable.

Throughout the tropics, forest departments are unable to protect forests. In most countries, their guards number in the hundreds or thousands, while forest inhabitants number in the millions. (See Table 1.) In Zaire, for instance, a forest department staff of about 800 is charged with protecting 100 million hectares inhabited by 15 million people. Such proportions are tanta-

TABLE 1.

Forest Area, Residents, and Forestry Staff, Selected Countries, Late Eighties

Country	Closed Tropical Forest Area	Estimated Forest Residents	Estimated Forest Department Staff
	(million hectares)	(million)	
Indonesia	114	15	17,000
Zaire	100	15	800
Peru	70	2	1,000
Papua New Guinea	36	3.5	400
Cameroon	16	0.4	800
Ecuador	14	2.2	800
Thailand	9	6	7,000

SOURCE: Nels Johnson and Bruce Cabarle, *Surviving the Cut: Natural Forest Management in the Humid Tropics* (Washington, D.C.: World Resources Institute, 1993).

mount to patrolling the state of New York with the police force of the city of Rochester.

The kind of joint management in force in Dolpura is a welcome step toward decentralized control of forests. Meanwhile, a few nations in the American tropics have taken more decisive strides. Under intense grassroots pressure, Bolivia, Brazil, Colombia, Ecuador, and Venezuela have all recognized land rights of tribes that have inhabited and conserved forests since time immemorial. In recent years, each of these countries has demarcated vast areas in the Amazon basin as indigenous homelands.[11]

The recognition of indigenous homelands in the Amazon has been the most hopeful sign for the world's forests in years. History will likely show that indigenous peoples and their supporters in human rights organizations saved more tropical forest than all the world's conservation groups put together. The forested homeland of the Yanomami in Brazil and Venezuela, for example, is as big as Uruguay, and far larger than any forest reserve in the world.[12] If African and Asian nations would follow this precedent, the prospects for tropical forests would improve markedly. Unfortunately, the rights of most traditional peoples are observed only in the breach. All but powerless in national politics, Asian tribes have minimal legal rights; African tribes have fewer. The pygmies of central Africa—the most ancient of all forest dwellers—hold no enforceable claims to the forests they have inhabited for 40,000 years.[13]

One step that outside organizations can take to help forest dwellers secure recognition of their rights is to assist them in developing accurate maps of their territorial boundaries. In most countries, maps of traditional tenure areas help bolster the claims of local communities. On the Atlantic Coast of Honduras, for instance, the Miskito Indians have recently completed a mapping project. Community leaders met to review scores of hand-drawn charts for accuracy before compiling them, with the help of trained cartographers, into a master map of the Miskito territory. After seeing them, even the Honduran military concluded that the Miskito Coast was not empty land ripe for settlement. Maps are not rights, but they help in the legal and political struggles to get them.[14]

The pygmies of central Africa—the most ancient of all forest dwellers—hold no enforceable claims to the forests they have inhabited for 40,000 years.

In the world's remaining temperate and boreal forests—concentrated in North America and Siberia—the social context of the forest economy is different, but the issue of tenure is equally important. In these regions, governments control public land not only in principle but in fact, because forest dwellers are fewer and governments better staffed. Where indigenous communities' rights go unrecognized, as in parts of Canada and Russia, securing them is as much a priority as it is in the tropics.

Elsewhere, the challenge is to establish tenure rules that encourage forest managers, whether private owners or government officials, to protect ecological integrity.

Forests in some industrial countries, particularly the United States, are less concentrated in the hands of the state than is the norm in the tropics, yet many laws still hamper good forest management. Most U.S. timber comes from private forestlands in the Southeast and Northwest, but small landowners in these regions are plagued by perverse incentives. For instance, inheritance tax laws often consider trees to be just another form of investment. Inheritance taxes can therefore turn the death of a family member into a must-cut situation: The children of a forest owner may have no choice but to liquidate their parents' groves to pay the tax bill.

As a shield against these forces, private and public institutions have created conservation easements—legally binding riders to land title deeds that forbid certain activities, such as clear-cut logging and urban development. Conservation easements ensure that lands cannot be taxed as if they were in commodity production. So far, however, conservation easements cover only a tiny fraction of forest lands.

Community land trusts offer another promising approach to encourage sustainable use of forests. In a land trust, a nonprofit board manages forests donated to it for the long-term benefit of the community. In Canada, provincial governments own virtually all forests, creating an obstacle to the development of land trusts. But even there, the logging and mining town of Revelstoke in British Columbia found an innovative solution: 20 citizens sued the owner of a tree farm license—effectively, a timber concession—for cut-and-run practices that violated the terms of the license. In late 1992, after winning in court, the citizens' Revelstoke Resource Development Committee borrowed enough money from commercial banks to buy the license. The committee is developing a multicentury plan for tourism, logging and collecting other forest products, and wood processing. Meanwhile, scores of volunteers are working in the forest to close logging roads and restore eroded stream banks.[15]

That tenure is a key determinant of the sustainability of forest economies is supported by reams of scholarly studies and economic analyses. But national and international efforts to save forests have focused primarily on such initiatives as planting more trees. It may seem heretical to belittle tree planting—now popularized as a virtuous or even heroic act—but the world's forests have never lacked regenerative capacity. What they lack is human allies. And they lack those allies because most tenure systems leave the day-to-day managers of forests without sufficient reason to protect them.

Rights to the Forests' Bounty

The need for property rights reform extends beyond forests' land and trees to the goods—from nuts and bamboo to organic chemicals and strands of DNA— increasingly extracted from wooded ecosystems. In fact, responsible property rights reform would extend also to the knowledge—or intellectual property—that local communities possess of these nontimber forest products. In the absence of tenure reforms, only entrepreneurs and corporations outside the forests will profit from these goods, diminishing the chances for sustainable livelihoods for those who live inside forests.

One recent case illustrates the potential of nontimber products, the state of the art in responsible corporate development of them, and at the same time, how far there remains to go in reforming tenure laws. In late 1991, Merck & Company, the world's largest pharmaceutical firm, sent a $1-million check to a small Costa Rican organization called the National Institute for Biodiversity (INBio). Merck also signed a contract that committed it to training Costa Rican researchers and paying INBio royalties from products it might develop in the future. In exchange, all INBio had to do was supply the pharmaceutical giant with a steady stream of leaves, bugs, and microscopic ooze from the country's tropical forests.[16]

Why would Merck enter such an agreement? To discover new medicines. In the eighties, researchers developed automated chemical screening techniques that allowed them to test compounds by the thousands for medicinal properties. The technology made it economical to begin exploring the great chemical factory of the natural world: tropical forests. In this stable, warm, and wet environment, plants, animals, insects, and microorganisms have evolved in millions of varieties, each equipped for the evolutionary battle with its own weaponry. Much of that weaponry is chemical, and some of it has already proved tremendously valuable to humans. Leech saliva, for example, contains a substance called hirudin that prevents clotting, allowing leeches to extract more blood from their victims. Inspired by the leech, modern medicine uses hirudin in skin transplants and to treat rheumatism, thrombosis, and contusions.[17]

The economic value of medicines taken from forests is staggering. Forty percent of prescription drugs dispensed by U.S. pharmacies have active ingredients derived from wild plants, animals, or microorganisms, many of them from forests. Given that the global pharmaceuticals industry is worth $200 billion, that the use of biologically derived medicines probably exceeds 40 percent of prescriptions outside the United States, and that a disproportionate share of biologically derived medicines originate in forest species, it is likely that annual sales of drugs with active ingredients derived from forests total more than $100 billion. Forests also provide most of the herbal medicine used by some 80 percent of developing-country residents.[18]

Furthermore, scientists have so far evaluated just a few thousand of the world's species, half of which dwell in tropical forests. Harvard University biologist Edward O. Wilson conveys the magnitude of the potential:

> A newly discovered species of roundworm might produce an antibiotic of extraordinary power, an unnamed moth a substance that blocks viruses in a manner never guessed by molecular biologists....An obscure herb could be the source of a sure-fire black-fly répellant—at last. Millions of years of testing by natural selection have made organisms chemists of superhuman skill, champions at defeating most of the kinds of biological problems that undermine human health.[19]

Shamans and other traditional healers have experimented with many of the species unknown to scientists. Unfortunately, their knowledge is disappearing even faster than are forest ecosystems, as the expanding world economy overruns and assimilates indigenous cultures.[20]

For Merck, signing the contract with INBio was a shrewd investment: a way to gain access to Costa Rica's estimated 12,000 species of plants and 300,000 species of insects. But it was also a principled decision, because it implicitly recognized that local parties—in this case the state—have property claims to untamed nature. Previously, groups ranging from pharmaceutical companies to crop breeders had considered plants, animals, and microorganisms free for the taking.

Forty percent of prescription drugs dispensed by U.S. pharmacies have active ingredients derived from wild plants, animals, or microorganisms, many of them from forests.

The agreement also had benefits for Costa Rican forest conservation: INBio designated one-tenth of the original Merck payment for the country's national parks budget, and similarly committed half of its royalties should Merck successfully develop products based on INBio's samples. Along the way, INBio will compile a national inventory of biological diversity and train a network of "para-taxonomists."[21]

But there is a caveat. By assuming that the living resources of Costa Rica's forests belong to the state, Merck and INBio ignored the possibility that the rightful owners of those forests are really the communities that dwell there. A year after the Merck-INBio agreement, the government of Costa Rica declared all wild plants and animals to be "national patrimony."[2] Thus, one of Latin America's most democratic and progressive states claimed ownership for itself of all the nation's wild plants and animals. This decree backs up the Merck agreement and effectively voids the claims of local citizens, some of whom had served as stewards of those living resources for generations.

Worse, neither Merck nor INBio stipulated that local lore

used to guide the selection of samples be treated as intellectual property belonging to the communities that developed it over long years of experimentation. The government of Costa Rica has performed no better, offering no protection under intellectual property law—the category that governs patents, copyrights, breeders' rights, and trade secrets—for forest dwellers' vast body of knowledge of the properties and uses of natural substances.

Around the world, not a single country extends intellectual property rights to indigenous ecological knowledge. If a traditional healer knows how to cure a skin disease with an herbal remedy, it is called folklore. If a pharmaceutical company isolates and markets the active chemical in the healer's herbs, it is called a medical breakthrough, protected with a patent, and rewarded with international monopoly power. Among biodiversity collectors, only the most ethical practitioners—notably the handful of ethnobotanists who study the plant uses of traditional cultures—adhere to codes of conduct that include negotiations with local communities to ensure that they benefit from any commercial products their knowledge helps create.

Under prevailing laws, then, forest dwellers in much of the world own neither their land nor their knowledge of that land. In the absence of such tenure, biological diversity prospecting—the "gene rush"—is likely to yield the same results as past resource booms in the tropics: more poverty, less forest.

The same is true of the other nontimber forest products, ranging from fruits and nuts to industrial fibers, that are now recognized as offering better livelihoods for many forest dwellers than timber extraction or slash-and-burn farming. In Belize, for instance, expert gatherers of forest products can earn between two and ten times as much per hectare as farmers who clear the forest for crops. Without secure control of these resources, however, their potential for sustainable employment will be lost.[23]

And that potential is still large. For example, Southeast Asian sales of rattan, the palm stems used to make wicker furniture, are worth $3 billion annually. These economic returns, moreover, are spread more equitably than timber earnings because of the labor intensity of harvesting them. Three times as many Indonesians work in the rattan industry as in the timber industry.[24]

Extractive reserves offer one promising contemporary model of local control over nontimber products. Owned by the government but managed by forest dwellers, extractive reserves were conceived in Brazil. Fourteen now exist there, covering close to one percent of the country's Amazon forests. Other extractive reserves have been created in Guatemala and Peru. In many ways, extractive reserves are a variation on joint forest management, distinct from the Asian form primarily in an emphasis on commercial harvesting of nontimber products rather than on subsistence uses of the forest. More important, both are actually just a new version of an ancient and pervasive form of collective land management.[25]

Under traditional law in much of the world, villages, clans, and other indigenous local institutions set aside broad areas for gathering forest goods. The Galik who conserved ironwood, for example, also reserved forests for collecting nuts, incense, fruit, and fuel. This latticework of local forest management systems still exists—though in a dilapidated, disempowered state—in much of its original range. No one knows what share of the globe's forests are managed as extractive reserves under this unofficial regime. But the best way to expand opportunities for nontimber forest products is to get nation-states to decentralize legal authority, returning at least a share of it to local, traditional, forest managers.[26]

The need for decentralization is as great in industrial countries as it is in the Third World, although the reasons are different. Collecting goods from the woods is a large, little-noticed industry, even in the world's richest nations. Karen Hobbs of Forks, Washington, for example, supplies florists with strange and beautiful mosses, lichens, and driftwood—all of it collected in the coastal temperate rain forests of the Pacific Northwest. Yet government managers see scant returns to their own budgets from sales of nontimber forest products. They consequently tend to favor timber, at the expense of low-impact forest products and services.[27]

Whether the resource is land, timber, floral supplies, or indigenous knowledge, secure tenure is the first necessary condition of a sustainable forest economy. Without it, the people who manage the world's forests have little reason to safeguard the forests' health.

Toward Ecological Pricing

Ecological pricing is the second necessary condition of a sustainable forest economy. Virgin timber is currently priced far below its full costs. For instance, the price of teak does not reflect the costs of flooding that rapacious teak logging has caused in Myanmar; nor does the price of old-growth fir from the U.S. Pacific Northwest include losses suffered by the fishing industry because logging destroys salmon habitat. Those losses are estimated at $2,150 per wild chinook salmon in the Columbia River, when future benefits to sports and commercial fishers are counted.[28]

Few attempts have been made to calculate the full ecological prices of forest products but they would undoubtedly be astronomical for some goods. A mature forest tree in India, for example, is worth $50,000, estimates the Center for Science and Environment in New Delhi. The full value of a hamburger produced on pasture cleared from rain forests is about $200, according to an exploratory study conducted at New York University's School of Business. These figures, of course, are speculative. Calculating them requires making assumptions about how many dollars, for instance, a species is worth—perhaps an imponderable question. But the alternative to trying—failing to reflect the loss of ecological functions at all in the price of wood and other forest products—ensures that the economy will continue to destroy forests.[29]

The full economic value of forest ecosystems is clearly huge. (See Table 2.) Forests provide a source of medicines worth billions of dollars, as described earlier. Their flood prevention, watershed stabilization, and fisheries protection functions are each worth billions more. Their scenic and recreational benefits also have billion-dollar values for both the world's growing nature tourism industry, and for local residents.[30]

Economists Ed Whitelaw and Ernest Niemi of Eugene, Oregon, for example, have demonstrated that the beauty of their state, including its mountains, forests, streams, and beaches, has substantial value to Oregonians. By comparing cost-of-living-adjusted pay scales nationwide in "footloose" industries—

TABLE 2.

Economic Services Provided by Intact Forest Ecosystems

Service	Economic Importance
Gene pool	Forests contain a diversity of species, habitats, and genes that is probably their most valuable asset; it is also the most difficult to measure. They provide the gene pool that can protect commercial plant strains against pests and changing conditions of climate and soil and can provide the raw material for breeding higher-yielding strains. The wild relatives of avocado, banana, cashew, cacao, cinnamon, coconut, coffee, grapefruit, lemon, paprika, oil palm, rubber, and vanilla—exports of which were worth almost $24 billion in 1991—are found in tropical forests.
Water	Forests absorb rainwater and release it gradually into streams, preventing flooding and extending water availability into dry months when it is most needed. Some 40 percent of Third World farmers depend on forested watersheds for water to irrigate crops or water livestock. In India, forests provide water regulation and flood control valued at $72 billion per year.
Watershed	Forests keep soil from eroding into rivers. Siltation of reservoirs costs the world economy about $6 billion per year in lost hydroelectricity and irrigation water.
Fisheries	Forests protect fisheries in rivers, lakes, estuaries, and coastal waters. Three fourths of fish sold in the markets of Manaus, Brazil, are nurtured in seasonally flooded *varzea* forests, where they feed on fruits and plants. The viability of 112 stocks of salmon and other fish in the Pacific Northwest depends on natural, old-growth forests; the region's salmon fishery is a $1-billion industry.
Climate	Forests stabilize climate. Tropical deforestation releases the greenhouse gases carbon dioxide, methane, and nitrous oxide, and accounts for 25 percent of the net warming effect of all greenhouse gas emissions. Replacing the carbon storage function of all tropical forests would cost an estimated $3.7 trillion—equal to the gross national product of Japan.
Recreation	Forests serve people directly for recreation. The U.S. Forest Service calculates that in eight of its nine administrative regions, the recreation, fish, wildlife, and other nonextractive benefits of national forests are more valuable than timber, grazing, mining, and other commodities.

SOURCES: Compiled by Worldwatch Institute from sources in endnote 30.

those that can be located anywhere—they have shown that people will work for less pay for companies in Oregon than in less beautiful locales. Oregonians, they argue, receive a "second paycheck" from nature worth about $500 apiece. Totalled statewide, these second paychecks almost equal the combined payrolls of all the state's lumber and wood-products firms.[31]

Forests also support local economies in ways that do not register fully in the money economy. Fruits, nuts, fibers, and wild game are mostly consumed in the subsistence economy of the poor, never entering the realm of cash transactions where they would show up in national accounting systems and count in the deliberations of those who make policy.

No one knows the full magnitude of the nontimber forest-products industries. Yet the harder researchers look, the larger they find them to be. Mater Engineering, Ltd., investigated markets for four relatively obscure plants—bear grass, huckleberries, salal, and sword fern—that grow in profusion in national forests near their offices in Corvallis, Oregon. They discovered that global sales of these four plants alone total $72 million a year. Likewise, they learned that foraging for wild mushrooms from the forests of the Pacific Northwest is an industry with sales in the hundreds of millions of dollars. The market value of nontimber forest products may exceed that of solid wood harvested from U.S. national forests—$1 billion in 1992.[32]

As buffers against climate change, forests' services may be worth trillions of dollars. By holding heat-trapping carbon dioxide out of the atmosphere, forests help prevent global warming and the rising sea levels and agricultural disruption that would accompany it. One hectare of Malaysian forest is estimated to provide carbon storage services alone worth more than $3,000 over the long-term.[33]

But forests' greatest value, and the most difficult to measure, is probably the diversity of life they contain. Forests possess most of the world's gene pool—the ultimate raw material for biotechnology, one of the world's emerging growth industries. And tropical forests hold vast potential for new crop plants as well. The fruits of the babassu palm, for example, yield more vegetable oil per hectare than any plant ever measured, yet so far

TABLE 3.

World Exports of Crops Descended from Tropical Forest Plants, 1991

Crop	Value
	(billion dollars)
Coffee	7.6
Citrus Fruits	3.8
Natural Rubber	3.4
Banana	3.1
Palm Oil	2.8
Cacao (for chocolate)	2.1
Pineapple	0.9
Vanilla	0.1
Total	23.8

SOURCE: values from U.N. Food and Agriculture Organization, *FAO Trade Yearbook, 1991* (Rome: 1993) crops from Nigel J.H. Smith et al., *Tropical Forests and Their Crops* (Ithaca, N.Y.: Cornell University Press, 1992).

babassu is all but unknown outside its wild range in the Brazilian Amazon.[3]

More immediately, forests harbor the wild relatives of dozens of crops: These related strains are crop breeders' first recourse in improving food and fiber crops, and protecting them against new pests and diseases. Export sales of crops that trace their origins to forests in the tropics and sub-tropics approached $24 billion in 1991. (See Table 3.) Among them are fruits including avocados, bananas, grapefruits, guavas, lemons, limes, mangos, oranges, papayas, and pineapples; spices including cinnamon, clove, and vanilla; nuts including Brazil nut, cashew, and macadamia; palm oil and rubber; and, perhaps most important

to the world's consumers, coffee and cacao—the source of choco-
late.

Most of these crops are currently grown from a narrow range
of genetic varieties, each vulnerable to catastrophic crop failures
if the wrong pest comes along. Virtually all the world's natural rubber, for example, is descended from seeds that the English
botanist Henry Wickham smuggled out of Brazil's eastern
Amazon in 1876. The world's natural rubber production, therefore, depends on just a few of the Amazon's hundreds of genetic varieties of rubber.[35]

Every attempt to grow rubber in plantations in the Amazon
has succumbed to the pests and blights that have evolved there
along with the plant. Southeast Asia's rubber plantations—
grown from Wickham's seeds—have so far escaped infestation;
they are the world's main source of the crop. How long they will
escape Amazonian pests, however, is anyone's guess. Their principal insurance is the diversity of wild rubber varieties still in the
Amazon. Each pest or blight that could menace rubber plantations probably has its antidote in the genetic traits of at least one
of those wild varieties.[36]

Similarly, most of the coffee grown in the Western Hemisphere
is descended from a single Javan seedling sent to Martinique in
1721. In 1970, coffee rust—a blight native to Africa—appeared in
Brazil, the world's leading coffee producer, and began spreading
northward through Central America and Mexico. Breeders racing
to locate resistant strains of coffee returned to coffee's evolutionary birthplace in the rapidly dwindling highland forests of
Ethiopia. There, they found coffee varieties resistant to as many
as 27 of the 33 known races of coffee rust.[37]

Ever since, researchers across Latin America have been cross-
breeding the resistant varieties with the commercial ones. They
have yet to win a decisive victory, but they have prevented a cat-
astrophe: When coffee rust reached Sri Lanka in 1869, it
destroyed that island's coffee industry in a matter of months.
Ironically, it was good fortune that the rust reached Brazil when
it did. Had it arrived twenty years later, the rust-resistant strains
might have vanished from Ethiopia, which has lost most of its
remaining forest cover since 1970.[38]

The full value of forests includes each of these components, from sources of medicines to pest controls. But, again, market prices count only the direct costs of extracting goods, not the full ecological costs. In accounting terms, the money economy is depleting its natural capital without recording that depreciation on its balance sheet. Consequently, annual losses come out looking like profits, and cash flow looks artificially healthy. For a business to do this—liquidate its plant and equipment and call the resulting revenue income—would be both self-destructive and, in many countries, illegal. For the money economy overall, however, self-destruction generally goes unquestioned.

How can we move toward ecological pricing? By changing government policies. A primary responsibility of governments is to correct the failures of the money economy, and global deforestation is surely a glaring one. Yet forest policies in most nations do the opposite: They accelerate forest loss. The first order of business for governments, therefore, is to stop subsidizing deforestation. The second is to use taxes, user fees, and tariffs to make ecological costs apparent in the money economy. Until the money economy is corrected in these ways, forest conservation will remain an uphill battle.

In tropical countries, most governments award timber concessions through a political process rather than competitive bidding, with the result that logs go at fire-sale prices.

Subsidies for deforestation are commonplace in both industrial and developing countries. In the United States, for example, the Forest Service, which long denied subsidizing logging, itself proposed in April 1993 that it would stop selling timber from 62 of the 156 national forests it administers because those forests had consistently lost money on timber sales. The announcement represented capitulation after a decade of pressure from conservation and taxpayer organizations, but it fell far short of ending the subsidized sale of logs from U.S. federal lands.[39]

Randal O'Toole of Cascade Holistic Economic Consultants in Oak Grove, Oregon, calculates that 109 national forests, accounting for 80 percent of national forest wood output, lost money on their timber sales in 1992. The loss to U.S. taxpayers was $499 million. The effect was to subsidize the logging of public forests, a disproportionate share of which are pristine and fragile high-country ecosystems, thereby shifting production away from more productive, already-disturbed lowland sites on private land. It also lowered the price of wood overall.[40]

In tropical countries, most governments award timber concessions through a political process rather than competitive bidding, with the result that logs go at fire-sale prices. In 1992, the Indonesian state took in just 17 percent of the "economic rent"— the earnings they could have made from sales—of timber concessions. Studies in other countries show that few capture more than half of what they could make.[41]

Once governments stop subsidizing destruction, they can begin to implement ecological pricing. The most powerful approach is to impose environmental taxes on goods produced at high ecological cost. Specifically, governments can shift the tax burden away from income and savings and toward "throughput"—the inputs of raw materials into the economy and the outputs of pollution and waste from it. For example, timber felled in intact, primary forests could be taxed at a high rate, timber from secondary forests at a lower rate; the lowest tax rate would be applied to timber certified by an accredited independent monitor as produced with the most sustainable practices available. Other deforestation taxes might be assessed on forest clearing for agriculture, real estate development, and road building.

Environmental taxes would need to be introduced across the board—not simply on forest products—or else businesses that use wood might substitute higher-impact commodities, such as aluminum and steel. Comprehensive ecological pricing would likely increase consumption of wood for certain uses, because the alternatives are sometimes even harder on the earth.

Enforcing environmental taxes on wood extraction would require a system for identifying and appraising wood's source.

Nongovernmental agencies are pioneering such a system. Now organized under the umbrella of the Forest Stewardship Council—an independent body of conservationists, human rights advocates, and forest products industries—wood certification has come about through a circuitous route.

In the early eighties, European citizens concerned about tropical deforestation began targeting the timber trade. They called for a boycott of the products of deforestation. By the early nineties, they had succeeded in convincing hundreds of local authorities in Austria, the Netherlands, Germany, and the United Kingdom to forbid the purchase of tropical timber for use in government offices. Then, in a crowning victory for the boycott promoters, B&Q, a leading U.K. chain of hardware stores, announced in December 1991 that it would phase out all products containing wood not harvested from sustainably managed forests. The company promised to be exclusively a "good wood" outlet by the end of 1995.[42]

User fees complement environmental taxation, by rewarding the managers of forests for all the services forests provide, rather than just for wood, minerals, and other commodities extracted at high ecological cost.

The trouble with the boycott of tropical timber was that it threatened to merely displace destructive logging practices from tropical forests to temperate and boreal ones. Eventually, the European activists and their North American counterparts realized the importance of certifying sustainably produced wood—whatever its region of origin—so that consumers had a clear choice. Supplying B&Q, and competitors that followed its lead, added urgency to that goal.

Since 1991, there have been a flurry of conferences, study missions, and task forces to develop standards of sustainability and methods for appraising logging methods. A few batches of wood have been officially certified. Yet there is little chance that

the bulk of the timber industry will be influenced by wood certification efforts, because most consumers will continue to use
whatever wood is cheapest. And, in the absence of generalized
ecological pricing, that means wood harvested without the pre-
cautions the certifiers require. What is promising about certification is that it is clearing the path for governments, sorting out
the questions of standards, methods, and enforcement that
could eventually make ecological taxes possible.

User fees complement environmental taxation, by rewarding
the managers of forests for all the services forests provide, rather
than just for wood, minerals, and other commodities extracted
at high ecological cost. In Colombia, for example, users of an
urban reservoir were assessed a small fee to pay for forest protection efforts in the watershed. Few such arrangements exist,
yet they are easy to imagine. Lowland irrigators could pay
upland forest communities for the water flowing in streams
through the dry season that those communities' forest stewardship ensures. Sports fishers and hunters could be charged fees
to help make habitat protection economically rewarding for
forest dwellers. Shippers and barge operators, who currently
pay for dredging through port fees, could pay for preventing river
sedimentation through upstream forest protection instead.
Floodplain developers could pay for the flood prevention services
that healthy forests provide, rather than for flood insurance.

Along these lines, economist O'Toole proposes a radical
redesign of the U.S. Forest Service, which he believes would better align forest management decisions with protection of forests'
full range of ecological services. Large organizations, O'Toole
argues, act first and foremost so as to maximize their budgets—
a thesis supported by research on bureaucratic behavior. The
Forest Service is no different. It favors commodity extraction over
recreational, nonconsumptive uses because peculiarities of its
legal structure allow it to retain most of the proceeds from timber sales. Its budget does not benefit, however, from camping,
fishing, hunting, hiking, skiing, or any other nonconsumptive
use. Nor does the Forest Service budget benefit from the water
its forests provide to farmers and cities downstream, nor from the
fish and wildlife that depend on its forests for habitat.[43]

O'Toole contends that the Forest Service will behave as a prudent manager of all the national forests' functions only when it can itself benefit from them. To do that, he recommends that the agency be reorganized to charge user fees for everything from camping to mining in national forests. The Service's own budget would come exclusively from those user fees. Because the nonconsumptive benefits of forests greatly exceed the consumptive ones in value, O'Toole believes the Forest Service would allow little timbering, mining, or other high-impact activities on the lands it controls. Charging visitors as little as $3 a day would generate more revenue than selling timber now does.[44]

The genetic wealth of forest ecosystems is more difficult to reflect in a system of user fees. One approach to doing so is to treat gene pools as a form of insurance, safeguarding cash crops against losses to pests and disease. The risks of catastrophic pest attacks or plant diseases are small for any particular crop in any given year, but over longer stretches of time they are all but certain. Since the genetic diversity of natural ecosystems, including forests, are agriculture's insurance against such disasters, perhaps industries that endanger it should pay to replenish it.

Specifically, resource extractors, land developers, and polluters could pay a share of the proceeds from ecosystem-damaging actions into regional biodiversity trust funds. Administered by committees of eminent biologists, those funds would be dedicated to rescuing the most endangered species, habitats, and genetic varieties through whatever method was most cost effective. Administrators might, for example, make grants to indigenous peoples' organizations struggling to secure land rights. They might purchase threatened habitats to create reserves or community land trusts. They might help timber towns diversify their economies. Or they might orchestrate debt-for-nature swaps.

Governments that aim for ecological pricing will need to implement environmental tariffs in addition to their environmental taxes and user fees. Otherwise, producers in full-cost countries would lose their domestic and international markets unfairly to producers in below-cost countries. Industrial countries could assess tariffs on beef from pastures in Central American rain forests and palm oil from plantations that replaced Southeast Asian forests, for example, while developing countries could impose tariffs on wood and paper from industrial countries that fail to stem the loss of forests to suburban sprawl. Unfortunately, these actions could violate the General Agreement on Tariffs and Trade, as currently written, pointing to the need to update this international accord to account for the environment.

In a global economy, where trade in wood products exceeded $98 billion in 1991, one region's unsustainable consumption sometimes comes at the expense of faraway forests. During the eighties, for example, loggers in California harvested about 4 billion board feet of timber from its forests each year—more than those forests could be expected to yield without risk to their ecological health. National and state-level ecological taxes and user fees on logging and forest clearing would be a highly effec-

FIGURE 1.
World Wood Consumption, 1950–91

Million cubic meters

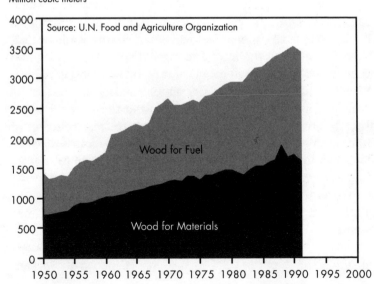

tive way for California to help sustain its own forests. But a thorny problem would remain. The state consumed about 10 billion board feet of timber each year during the eighties—drawing on other regions of the United States and the world for wood it did not grow itself. In the absence of universal ecological pricing, the state's consumption would go undiminished. The net effect would be to export—rather than halt—deforestation.[45]

Taming Consumption, Greening Production

Getting forest prices to tell the ecological truth would help move the forest economy toward sustainability in three principal ways: by encouraging more efficient and less wasteful uses of wood; by fostering more sustainable substitutes for wood; and by encouraging ecologically sound forest management.

Artificially low, nature-blind prices promote inefficiency, particularly in the use of wood, the forests' major commodity product. In 1991, the world economy consumed 3.4 billion cubic meters of wood. (See Figure 1.) That amount is 2.5 times as much wood as was used in 1950—one third more per person. Consumption rates per person remain far higher in industrial nations than in developing ones, but recent growth in total and per capita consumption has taken place mainly in the Third World.[46]

Half the world's wood is burned as fuel, mostly in developing countries; the other half is cut as timber and milled into boards, plywood, veneer, chipboard, paper, paperboard, and other products, mostly in industrial countries. (The gathering of fuelwood is a hardship for Third World women and children and taxes open woodland ecosystems; it rarely threatens closed-canopy forests, however, and is therefore omitted from this paper.)

The conventional wisdom among foresters counsels that "people demand wood." Much as energy producers once believed their product indispensable to industrial production, most timber producers treat wood as an end in itself, and assume demand for it is bound to grow. In the words of a 1974 public relations brochure published by the U.S. Forest Service, "We harvest timber because it is needed for man's survival." The U.N. Food and Agriculture Organization and the U.S. Forest Service have both consistently overestimated future demand for wood, based largely on this assumption.[47]

In fact, per capita wood consumption has been declining in industrial nations for most of a century, first as fossil fuels replaced fuelwood, and later as lumber was both used more efficiently and replaced with metal, plastics, and other substitutes. For example, chipboard—fibers of wood waste pressed and glued together—has replaced lumber in a substantial share of furniture manufacturing worldwide. This opens new supply options: Fiber can come from recycling, from tree species such as alder that were previously regarded as weeds in temperate zones, from thousands of little-known species currently bulldozed aside to reach mahogany and teak in tropical forests, and from fast-growing non-tree crops such as kenaf. Under comprehensive ecological pricing, wood might be substituted for metals and plastics in some applications, if the latter had higher ecological costs, but efficiency improvements would probably still reduce wood consumption overall.

In the energy industry, many have now learned that energy consumption is sensitive to price, and that ample opportunities exist for cost-effective savings through efficiency improvements. Some now profit from these opportunities. Most of the wood products industry, however, remains stuck in the old commodity mindset, focusing on quantity over quality, volume over value. Yet there is abundant evidence that people do not demand wood. They demand houses, tables, chairs, printed information, and other useful amenities. Wood is simply a means to an end—a means that will be used more or less frugally depending on its price.

Ecological pricing would discourage the waste that is rife in the world's consumer societies. A principal use of solid lumber around the world is for construction of new houses; in North

America, it is the leading use. Already, some manufacturers of housing components and furniture are recognizing the wood savings they can achieve. Andersen Window, the world's largest window maker, plans to collect old sashes in the trucks they use to distribute their windows. They will remanufacture the recycled sash into new, better insulated products. TrusJoist International's I-beams—high-strength structural lumber manufactured from wood wastes and second-growth timber—are quickly replacing solid, old-growth lumber for roof framing in North American home building. And Herman-Miller, Inc., plans to begin collecting, repairing, refinishing, and reselling used furniture from its product line.[48]

The scale of the wood-efficiency potential stands out clearly in the home Steve Loken of Missoula, Montana, built in 1991. Loken sought opportunities to save wood—and other natural resources—at every step. His house is constructed to high specifications of quality, appearance, and safety, but it required

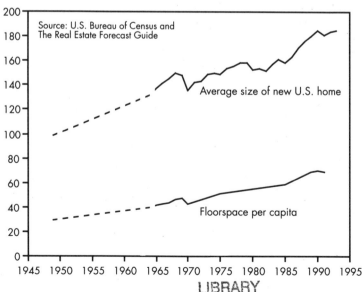

FIGURE 2.

Residential Space per Capita and Average Size of U.S. Homes, 1949–93

Square meters

Source: U.S. Bureau of Census and The Real Estate Forecast Guide

Average size of new U.S. home

Floorspace per capita

just one-fifth as much solid wood as a typical house of its size. For homebuilders, ecological pricing would turn Loken's house from a curiosity into a model. For consumers, it would divulge the costs of their choices.[49]

At present, unfortunately, housing trends have worked strongly against sustainability, especially in North America, where house size has ballooned. In 1949, the average new house in the United States was 100 square meters; by 1993, it was 185 square meters. (See Figure 2.) As houses have grown, the families that buy them have shrunk. The number of people sharing each American residence has fallen by almost half since 1900, from 4.8 to 2.6 people per residence. Partly, that is because parents are having fewer children—a good thing for the environment—but mostly it is because fewer extended families are

FIGURE 3.

World Paper Consumption, 1913–91 and Recycled Paper Consumption, 1983–91

Million tons

Source: U.N. Food and Agriculture Organization

living together.[50]

The combined effect of larger houses and fewer inhabitants is more space—and more lumber use—per person. Since 1950, residential floorspace per capita has risen from 29 to 69 square meters in the United States. On average, North Americans have half again as much residential space per person as West Europeans, nearly twice as much as Japanese, and more than three times as much as Russians. The Japanese, however, are racing to catch up. Despite a stable population, the country has added between 200 and 300 million square meters of residential space each year for the past decade; at least 75 million square meters of the annual addition has been built with wood, most of it imported from forests around the Pacific Rim. Ecological pricing would help people select a dwelling size that not only they but the planet can afford.[51]

Ecological pricing would also expose paper markets to a healthy dose of ecological reality. Per capita paper consumption is soaring worldwide, a disheartening trend for the environment because the pulp and paper sector is the biggest polluter

TABLE 4.
Paper Consumption per Capita, 1991

Region/Country	Consumption
	(kilograms paper and paperboard)
North America	302
Japan	228
Europe	163
Latin America	26
Asia	20
Africa	6

SOURCE: Pulp and Paper International, *1992 North American Fact Book, Pulp and Paper* (San Francisco: 1991).

among wood products industries. It is a heavy user of energy and water, and a major source of toxic water pollutants such as dioxin. Paper production has grown rapidly throughout the twentieth century, increasing its output twentyfold since 1913. (See Figure 3.)[52]

Rising literacy around the world accounts for some of this growth. But the largest share of paper and paperboard ends up in industrial countries as boxes and other packaging—much of it superfluous—and as a medium for the dissemination of advertisements.

In consumer societies, advertising bloats the daily mail: 14 billion glossy, difficult-to-recycle mail-order catalogs, plus 38 billion other assorted ads clog the post office each year in the United States. Most of those items—98 percent of direct-mail advertising letters, according to the marketing journal *American Demographics*—go straight into the trash. Advertisements also fill periodicals: Most American magazines reserve 60 percent of their pages for advertising, and some devote far more. Advertising typically comprises 65 percent of newspapers in the United States, up from 40 percent half a century ago. European and Japanese periodicals feature fewer ads; Europe and Japan also boast more conscientious recycling and more modest packaging, all of which contribute to their lower paper consumption rates. (See Table 4.)[53]

Ironically, office automation has also driven growth of paper consumption. As of 1992, for example, the world had more than 19 million photocopiers, devices only introduced in 1948. Since 1955, world consumption of printing and writing paper, including photocopying paper, has increased sevenfold. In the United States alone, the spread of office printers, photocopiers, and fax machines spurred a near doubling of office paper consumption during the eighties.[54]

Fortunately, recycling rates are rising in many of the major paper-consuming nations. From 1983 to 1991, the share of all paper and paperboard worldwide manufactured from recycled fibers rose from 30 to 37 percent. Still, recycling has yet to dent the world's appetite for virgin-fiber pulp, merely slowing its growth. As with wood, the efficiency of paper use would improve under ecological pricing. Newspapers might shift their classified advertising sections onto electronic databases available by telephone. Packagers would trim the excess. Catalog distributors would better hone their mailings. Wholesalers would replace corrugated boxes with reusable shipping crates. The long-awaited paperless office might finally arrive.[55]

Under a true accounting system, clear-cut timber would be far more expensive than wood from careful, low-impact logging. In current market prices, however, recklessness appears cheap. That illusion is why just one-tenth of one percent of all tropical wood is logged on a "sustained-yield" basis: one in which annual harvest does not exceed annual regrowth. Worse, even so-called sustained-yield timbering has proved unsustainable in the long run, because it is often accomplished by growing trees in a monoculture, instead of maintaining the vibrant ecosystems and diversity of species that natural forests support.[56]

Per capita paper consumption is soaring worldwide, a disheartening trend for the environment because the pulp and paper sector is the biggest polluter among wood products industries.

If prices did equal full cost, however, techniques like the "new forestry" emerging in North America, the world's largest timber producer, would spread swiftly. Originating in the U.S. Northwest, one of the few remaining temperate regions to possess old-growth forests of any size, new forestry strives to minimize the damage caused by logging. It aims to leave the forest's ecological functions unimpaired and its biological diversity undiminished. New forestry principles include preserving the forest's ecological structure—its canopy layers, its patterns of plant succession, and its cycling of nutrients—and setting aside broad undisturbed corridors around watercourses. New foresters generally minimize road building, and often use management techniques that mimic natural forces that disturb forests, such as fire and wind.

TABLE 5.

Production of Industrial Roundwood, Top 20 Nations and World, 1991

Country	Industrial Roundwood	Share of Total
	(million cubic meters)	(percent)[1]
United States	410	26
Soviet Union	274	17
Canada	171	11
China	90	6
Brazil	74	5
Sweden	47	3
Germany	40	3
Malaysia	40	3
France	34	2
Finland	31	2
Indonesia	29	2
Japan	28	2
India	25	2
Australia	18	1
Spain	15	1
Czechoslovakia	15	1
Poland	14	1
New Zealand	14	1
Austria	14	1
South Africa	13	1
Others	200	13
World	1,599	100

SOURCE: U.N. Food and Agriculture Organization, *Forest Products Yearbook, 1991* (Rome: 1993).

[1]Does not total 100 due to rounding.

New forestry is still developing, and it varies in application from one ecosystem to the next. But in the bioregion where it was developed, despite its demonstrated ecological superiority to conventional clear-cutting, only the intervention of the nation's president ensured its use on federal timber lands.

What would it take to spread new forestry principles quickly? How can loggers on both public and private land throughout North America be persuaded to change? What about the timber businesses of the former Soviet Union, the world's second largest wood producer? And all the other timber extracting countries? (See Table 5.) The answer is ecological pricing, implemented nationally and reinforced internationally through tariffs, as described above.

A Question of Power

More difficult than designing tenure and price policies is political change, the third necessary condition of a sustainable forest economy. Past reform attempts have mainly foundered. The world's forest economy functions—or malfunctions—as it does because its current structure benefits powerful groups. Such groups can be expected to fight tenaciously in defense of their privileges. Overcoming their concentrated economic and political power will require concerted campaigns, tireless grassroots organization, and ingenious political strategies. But unless the disenfranchised who depend on forests for sustenance gain greater influence over the fate of forests, there is little hope for saving the forests. In particular, unless the viselike grip of big timber interests—and miners, ranchers, and related resource extractors—can be broken, all bets for forest conservation are off.

Malaysia is the extreme case and illustrates the point clearly. The largest exporter of tropical timber in the world, Malaysia is stripping its Borneo provinces of trees at a breakneck pace. Work in some operations continues around the clock, with gigantic floodlights illuminating the forests. The ruin of local ecosystems, devastation of indigenous homelands, and economic folly of mining a potentially renewable resource—all have been thoroughly documented and publicized in Malaysia and abroad.[57]

Why should the government of Malaysia continue on this path? Logging in Malaysian Borneo is driven by the collusion of power and money. By tradition, elected leaders in Malaysian provinces have authority to distribute contracts to exploit public resources, notably timber. This prerogative has become a crucial part of their power base. Officials distribute logging concessions to loyal supporters who level the trees for quick profits— estimated in the hundreds of millions of dollars. A share of the proceeds helps keep the logger's patron in office, and may make him a millionaire as well.[58]

To varying degrees, this bond between timber money and political power is found in all the world's major timber economies. Indonesian timber magnate Prajogo Pangestu, who owns concession rights and wood products industries worth an estimated $5 billion, continues to expand his control of the nation's forests with the help of those at the highest levels of government. Similarly, the Philippine Congress is packed with loggers and members of logging families.[59]

Garden-variety corruption is also widespread in the disposition of the world's forests. Papua New Guinea appointed a commission to investigate the enforcement of national forestry laws, particularly on foreign timber companies, in New Ireland Province. The commission reported that foreign timber companies were "roaming the countryside with the self-assurance of robber barons; bribing politicians and leaders, creating social disharmony and ignoring laws in order to gain access to, rip out, and export the last remnants of the province's valuable timber."[60]

In Indonesia, a study conducted by the environmental organization Walhi found that concessionaires commonly bribe forest inspectors so they will not enforce forestry rules. Just 22 of 578 concessionaires followed the rules, and 70 percent of the timber harvest was taken illegally in some provinces. In the Philippines, the illegal timber trade may amount to four times the size of the legal trade.[61]

The political influence of timber money is not confined to

the Third World. In the U.S. timber states of Washington and Oregon, the wood products industry outspent environmentalists six-to-one in contributions to congressional candidates between 1985 and 1992; during those years, grateful members of Congress from these states set higher logging targets for national forests in their districts than the Forest Service itself recommended.[62]

Challenging the collusion of power and money is, in the best cases, a monumental task. In the United States, one of the most democratic societies in the world, it took five years of a nationwide grassroots campaign and a court injunction to arrest the clear-cutting of primary forests in the Pacific Northwest—even though such logging was patently illegal under both the National Forest Management Act and the Endangered Species Act.[63]

In less democratic societies, those who question the prerogatives of economic power all too often end up as murder statistics in human rights reports. On July 17, 1993, to cite just one case, three men hacked to death Filipino development worker William Rom—a researcher documenting the economic and educational needs of the Mamanua tribal people. His apparent transgression was to work for a nongovernmental organization that exposed illegal logging on tribal lands.[64]

In the Philippines, the illegal timber trade may amount to four times the size of the legal trade.

Again, Linka Ansulang speaks for such unsung heroes. An aged tribal woman from Mindanao—the same island where Rom was killed—Ansulang led 31 Manubo families in reclaiming tribal forestland from state corporations. Long ago, the government made a verbal agreement with her people to provide farm tools and employment in exchange for a 35-year lease of the tract. Although the state never fulfilled its promises, the Manubo left the land for 35 years. When the contract expired—on May 25, 1990—they returned, greeted by a local mayor's threats to bury them there. The standoff continues. Through it all, Ansulang and her people have remained committed to a peaceful reoccupation of what is rightfully theirs. "All we have to defend ourselves with," she says with a smile, "are carabao [water buffalo] and grass."[65]

Pressure for change from citizens such as Rom and Ansulang is finding a new complement in pressure from consumers far afield. In April 1993, for example, the government of Indonesia bought large advertisements in influential American newspapers touting the ecological sensitivity of its forest policies. "Forests Forever," proclaimed the ads. About the same time, the Canadian government launched a $6-million public relations effort in Europe, aiming to polish an image tarnished by media coverage of clear-cutting on the Pacific Coast. The premier of British Columbia vowed to "battle the campaign of misinformation bombarding our international customers."[66]

In fact, neither Indonesia nor Canada has much to brag about. Rates of primary forest clearing are high in both countries. Indonesia certainly will not have "forests forever" if it continues as it has—trampling on the rights of forest dwellers, serving up national resources to an oligopoly of the president's cronies, and doing it all at fire-sale prices. By the same token, some Canadian provinces have yet to recognize indigenous peoples' land rights. Many provinces subsidize timber extraction. And most cater to politically powerful logging, pulp, and paper industries.

Still, these publicity campaigns are backhanded compliments to the advocates of sustainability. That the Indonesian and Canadian governments would feel it necessary to defend their forest policies says that consumer perceptions of environmental stewardship matter. They do not yet matter enough to change government policy and redesign the forest economy. But they matter. The challenge is to make them matter more.

Old Forests, New Jobs

The main argument that entrenched interests make against sustaining forests is that if the trees stand, jobs will be lost. In the U.S. Northwest, the question of whether to protect the ecological integrity of old-growth forests was popularized as a conflict between jobs and the endangered northern spotted owl. Rank-and-file loggers became the most politically effective voices arguing on behalf of business as usual. Behind the scenes, however, the political muscle was in the timber companies, and they fought for profits, not jobs. The companies had been quietly trimming payrolls for decades as advancing technologies made labor-saving possible.

Still, the opponents of sustainable economics commonly use the threat of job loss to derail reform—arguing in essence that society cannot afford to protect ecological health if doing so involves displacing workers. The logic is spurious, since it assumes that work could continue as ecosystems unravel. But to counter such an argument is nevertheless essential, especially since real reform will doubtless entail social turbulence.

Like the whalers denied a livelihood by the international whaling moratorium, or the weapons manufacturers put out of work by the end of the Cold War, the loggers, developers, and other workers employed by the industries of deforestation have a wrenching period of transition ahead of them. Most are dedicated to their communities, skilled in their occupations, and proud of their work. Few of them get rich from clearing forests, and many barely keep hearth and home together. They deserve the support of their compatriots as they make the change to sustainable livelihoods. Their jobs, however, are no more a reason to continue deforestation than jobs in weapons plants are a reason to go to war.

If the military analogy is apt, then the answer for these workers is economic conversion—the process through which people and communities make the transition to sources of livelihoods they can depend on. Assistance can come from the state or from public-interested organizations such as charities and unions. On the Olympic Peninsula of Washington State, for example, the nonprofit group Woodnet helps would-be entrepreneurs in declining timber towns to develop markets for their high-quality handiworks—durable wood products such as chess boards, musical instruments, children's toys, and cabinets.

Hundreds of similar, small-scale efforts are emerging in the resource dependent towns of the world's forest zones, demonstrating that the future for these communities no longer lies in extracting large volumes of timber, minerals, and other raw materials. It lies in adding value to them by manufacturing finished products. In the United States, for example, logging 1 million board feet of timber yields about 3 jobs. Milling it into lumber generates 20 jobs. And crafting it into furniture creates 80 jobs. With each step up the processing chain, each job requires less stuff and more skill. The earth suffers less, and livelihoods become more secure.[67]

Markets do eventually achieve conversion unaided, but at a high cost in broken homes and shattered lives. Smoothing the conversion process—through measures such as unemployment insurance, worker retraining, and community development banks—is a legitimate function of the state. To fund such programs, policymakers could tap a share of the proceeds from environmental taxes and user fees implemented to put the forest economy on a sound footing in the first place. The ultimate challenge is to make ecological services sufficiently remunerative for all forest communities—indigenous tribes and Third World villagers; logging towns and forest agencies in both North and South—that they act as defenders of forests.

Aiming for Permanence

In their disappearance or regeneration, forests serve as an indicator of progress toward the paramount goal of reconciling the human enterprise with the Earth's biological design. Bill Dietrich, science correspondent for the *Seattle Times*, concluded as much after chronicling the five-year controversy that raged over the protection of old-growth forests in the U.S. Northwest.

The conflict can be seen as a test case for the world. Would the richest nation in the history of the planet spare the last 5 percent of its natural forests from logging and voluntarily make a transition away from timber mining before wood scarcity forced it to do so? (As of late 1993, the answer seemed to be "yes"—a good portent.) As the controversy reached its climax in a presidential summit in April 1993, Dietrich characterized the issue: It is certainly not just a fight over owls. And despite the staggering numbers one can cite—667 species now identified as tied to old-growth forest, 214 fish runs threatened across three states, 463 overcut watersheds identified in western Washington alone, 75 percent of the region's rivers suffering environmental damage—it is not even just about trees, or murrelets, or salmon, or clear-cuts. It is about our most basic values.... It is about whether this region can remain a place of special grandeur, character and optimism... or whether it is losing its sense of itself as blessed and promising. Whether God's Country is being homogenized into the ordinary: our cities a standardized U.S. chic, our villages gentrified and strip-malled, our acreage put to the narrowest economic measure.[68]

Dietrich is right. The question of forests is as much about ourselves as it is about lumber, water, and biological diversity. It is about what we are and about what kind of society we want to be. On the evolutionary stage, are we tragic or transcendent characters? Planetary locusts, or planetary stewards?

In the end, the fate of the earth's wooded lands is tied to the fate of their inhabitants. Either rights to ancestral lands will be defended with the full force of the law, or the forest will fall. Either forest dwellers will gain a share of the economic worth of the ecological services their forests provide, or the forests will fall. Either forest residents will be allowed into the corridors of power, or the forests will fall. Other things can help save forests, but these are fundamental: property rights, price, and power. As they are currently structured, the epoch of sweeping deforestation that has gripped the world since 1950 will only continue.

If those who manage forests can benefit from sustaining them in the full vigor of their ecological functioning—through reforms in tenure, price, and power—the future need not echo the recent past. We can envision a world of regenerated forests and healthy economies: one where forest dwellers, with their intimate practical knowledge of the woods, join forces with ecologists to plan the use of their lands. We can hope for a landscape designed for ecological health and economic productivity. We can expect an economy that thrives on value rather than volume, rewards thrift, and aims for permanence. And we can imagine a time-lapse film of the earth from space, showing forests gradually growing thicker, spreading outward, returning to health.

Notes

1. Global forest cover from John F. Richards, "Land Transformation," in B.L. Turner II et al., eds., *Earth as Transformed by Human Action* (New York: Cambridge University Press, 1990); and from R. A. Houghton et al., "Changes in the Carbon Content of Terrestrial Biota and Soils Between 1860 and 1980: A Net Release of CO_2 to the Atmosphere," *Ecological Monographs*, September 1983; forest history throughout introduction primarily from Alexander S. Mather, *Global Forest Resources* (Portland, Oregon: Timber Press, 1990).

2. Forest cover estimates from Richards et al., op. cit. note 1, from Houghton et al., op. cit. note 1, and from Sandra Postel, "Carrying Capacity: Earth's Bottom Line," in Lester R. Brown et al., *State of the World 1994* (New York: W.W. Norton & Co., 1994).

3. Tree cover from Postel, op. cit. note 2; intact forest ecosystems from Sandra Postel and John C. Ryan, "Reforming Forestry," in Lester R. Brown et al., *State of the World 1991* (New York: W.W. Norton & Co., 1991).

4. John C. Ryan, *Life Support: Conserving Biological Diversity*, Worldwatch Paper 108 (Washington, D.C.: Worldwatch Institute, April 1992)

5. Tropical Forestry Action Plan, *Report of the Independent Review*, Kuala Lumpur, Malaysia, May 1990; Marcus Colchester, "The International Tropical Timber Organization: Kill or Cure for the Rainforest?" *The Ecologist*, September/October 1990; deforestation trends from Postel, op. cit. note 2.

6. Jagat Mehta, president, Seva Mandir, Udaipur, Rajasthan, India, private communication, July 24, 1991; Seva Mandir's efforts from author's observations during visit, late July 1991.

7. India from Betsy McGean, forestry consultant, Chevy Chase, Md., private communication, October 12, 1993; Philippines from Delfin J. Ganapin, Jr., deputy secretary, Department of Environment and Natural Resources, Quezon City, Philippines, private communication, July 14, 1992.

8. Nationalization of tropical forests from Theodore Panaytov and Peter S. Ashton, *Not by Timber Alone* (Washington, D.C.: Island Press, 1992).

9. Linka Ansulang, Carmen, Philippines, private communication, July 9, 1992.

10. Nancy Lee Peluso, "The Ironwood Problem: (Mis)Management and Development of an Extractive Rainforest Product," *Conservation Biology*, June 1992.

11. Alan Thein Durning, *Guardians of the Land: Indigenous Peoples and the Health of the Earth*, Worldwatch Paper 112 (Washington, D.C.: Worldwatch Institute, December 1992).

12. Area of Yanomami reserves from ibid.; sizes of parks and nature reserves from International Union for Conservation of Nature and Natural Resources, *1990 United Nations List of National Parks and Protected Areas* (Gland, Switzerland, and Cambridge, U.K.: 1990).

13. Kirk Talbott, "Central Africa's Forests: The Second Greatest Forest System on Earth," World Resources Institute, Washington, D.C., January 1993.

14. Mac Chapin, program director, Rights and Resources, Arlington, Virginia, private communication, June 3, 1993.

15. Scott Zens, research assistant, College of Forest Resources, University of Washington, Seattle, Washington, private communication, September 21, 1993.

16. Walter V. Reid et al., *Biodiversity Prospecting* (Washington, D.C.: World Resources Institute, 1993).

17. Hirudin from Edward O. Wilson, *The Diversity of Life* (Cambridge, Mass.: Harvard University Press, 1992).

18. Share of U.S. prescriptions from World Resources Institute, World Conservation Union, and United Nations Environment Programme, *Global Biodiversity Strategy* (Washington, D.C.: 1992); value of world pharmaceuticals industry and share of developing country residents employing herbal medicine from Reid et al., op. cit. note 16.

19. Wilson, op. cit. note 17.

20. Loss of indigenous cultures from Durning, op. cit. note 11.

21. Reid et al., op. cit. note 16.

22. Ibid.

23. Michael J. Balick and Robert Mendelsohn, "Assessing the Economic Value of Traditional Medicines from Tropical Rain Forests," *Conservation Biology*, March 1992.

24. Rattan from Jenne H. De Beer and Melanie J. McDermott, *The Economic Value of Non-timber Forest Products in Southeast Asia* (Amsterdam: Netherlands Committee for IUCN, 1989).

25. Ryan, op. cit. note 4.

26. Galik from Peluso, op. cit. note 10.

27. William Dietrich, *The Final Forest* (New York: Penguin, 1992).

28. Net present value of salmon from Carolyn Alkire, *Wild Salmon as Natural Capital* (Washington, D.C.: The Wilderness Society, 1993).

29. Thomas Gladwin, Leonard N. Stern School of Business, New York University, New York, private communication, September 5, 1993; Center for Science and Environment, *The Price of Forests* (New Delhi: 1993.)

30. Table 1 based on the following: export value of commodities that originated in tropical forests from U.N. Food and Agriculture Organization (FAO), *FAO Trade Yearbook 1991* (Rome: 1993); share of Third World farmers depending on forests for water from World Bank, *Wildlands* (Washington, D.C.: 1987); flood control value from Panayotov and Ashton, op. cit. note 8; cost of siltation from K. Mahmood, *Reservoir Sedimentation: Impact, Extent and Mitigation* (Washington, D.C.: World Bank, 1987); Manaus fisheries from Michael Goulding, *The Fishes*

and the Forest (Berkeley, California: University of California Press, 1980); Northwest fisheries from Jack Ward Thomas et al., Viability Assessments and Management Considerations for Species Associated with Late-Successional and Old-Growth Forests of the Pacific Northwest (Washington,D.C.: Forest Service, U.S. Department of Agriculture, 1993); tropical deforestation's responsibility for climate change from Richard A. Houghton, "The Role of the World's Forests in Global Warming," in Kilaparti Ramakrishna and George M. Woodwell, eds., World Forests for the Future (New Haven, Connecticut: Yale University Press, 1993); replacement cost of forests' carbon storage function from Panayotov and Ashton, op. cit. note 8; U.S. Forest Service from Randal O'Toole, Reforming the Forest Service (Washington, D.C.: Island Press, 1988).

31. W. Ed Whitelaw and Ernest G. Niemi, "Money: the Greening of the Economy," Old Oregon, Spring 1989.

32. Value of four products and of mushroom industry from Mater Engineering, Ltd., "Analysis and Development of a Conceptual Business Plan for Establishing a Special Forest Products Processing Plant," prepared for Sweet Home Ranger District, Sweet Home, Oregon, June 30, 1992; estimated value of non-timber forest products from Catherine Mater, vice president, Mater Engineering, Ltd., Corvallis, Oreg., private communication, September 7, 1993; 1992 value of solid wood from national forests from Randal O'Toole, "1992 TSPIRS Recalculations," Forest Watch (Cascade Holistic Economic Consultants, Oak Grove, Oregon), April/May 1993.

33. Panayotov and Ashton, op. cit. note 8.

34. Babassu from Wilson, op. cit. note 17.

35. Richard Evans Schultes, "Ethnobotany, Biological Diversity, and the Amazonian Indians," Environmental Conservation, Summer 1992.

36. Ibid.

37. Nigel J.H. Smith et al., Tropical Forests and Their Crops (Ithaca, N.Y.: Cornell University Press, 1992).

38. Ibid.

39. Keith Schneider, "U.S. Would End Cutting of Trees in Many Forests," New York Times, April 30, 1993.

40. O'Toole, op. cit. note 32.

41. Victor Mallet, "Rules Must Be Right—And Upheld," Financial Times, May 13, 1993; Robert Repetto and Malcolm Gillis, eds., Public Policies and the Misuse of Forest Resources (New York: Cambridge University Press, 1988).

42. B&Q from Alan Knight, "B&Q's Timber Policy towards 1995," B&Q plc., Eastleigh, U.K., December 1992; local governments from Nels Johnson and Bruce Cabarle, Surviving the Cut: Natural Forest Management in the Humid Tropics (Washington, D.C.: World Resources Institute, 1993).

43. O'Toole, op. cit. note 30.

44. Ibid.

45. World exports from FAO, *1991 Forest Products Yearbook* (Rome: 1993); California forestry from World Resources Institute, World Conservation Union, and United Nations Environment Programme, op. cit. note 18.

46. FAO, *Forest Product Yearbook 1960* through *1991* (Rome: 1962 through 1993). Figure 1 from ibid. "Wood for fuel" is fuelwood and charcoal; "Wood for materials" is industrial roundwood.

47. U.S. Department of Agriculture, Forest Service, *Patience and Patchcuts* (San Francisco, California: 1974).

48. Loren Abraham, Andersen Windows, speech at First North American Conference on Trade in Sustainable Forest Products, Washington, D.C., May 26, 1993; TrusJoist International from Dori Jones Yang, "A Lumberman Goes Against the Grain," *Business Week*, March 29, 1993; Bob Johnston, Herman-Miller, Inc., speech at First North American Conference on Trade in Sustainable Forest Products, Washington, D.C., May 26, 1993.

49. Julie Titone, "Tomorrow's House," *American Forests*, March/April 1993.

50. Figure 2 from following: new house size from Michael Sumichrast, *The Real Estate Forecast Guide* (Gaithersburg, MD.: 1992), with historical information from National Association of Home Builders (NAHB), Washington D.C., private communication, August 28, 1993; residential floorspace per capita from Department of Commerce, Bureau of the Census, *Statistical Abstract of the United States 1992* (Washington D.C.: 1992) and *Bicentennial Edition, Historical Statistics of the United States, Colonial Times to 1970* (Washington, D.C.: 1975); people per U.S. residence from Bonnie Maas Morrison, "Ninety Years of U.S. Household Energy History: A Quantitative Update," *1992 Summer Study* of the American Council for an Energy-Efficient Economy, (Washington, D.C.: 1992).

51. U.S. floorspace from Department of Commerce, Bureau of the Census, op. cit. note 50; international comparisons of residential floorspace from Lee Schipper, staff senior scientist, Lawrence Berkeley Laboratory, Berkeley, California, private communication, November 6, 1991; Japanese floorspace from International Tropical Timber Organization, *Annual Review and Assessment of the World Tropical Timber Situation, 1990-1991* (Yokohama, Japan: 1992).

52. Figure 3 based on Pulp and Paper International, *PPI's International Fact and Price Book* (Brussels: 1993), and on FAO, op. cit. note 46, on FAO, *European Timber Statistics 1913-1950* (Geneva: 1953), and on FAO, *World Forest Products Statistics 1946-1955* (Rome: 1957)

53. Alan Thein Durning, "Can't Live Without It," *World Watch*, May/June 1993.

54. Photocopiers from Lynn Ritter, senior industry analyst, Dataquest Inc., San Jose, Calif., private communication, May 28, 1993; year of introduction of photocopiers from Xerox Corporation, *1993 Xerox Fact Book* (Stamford, Connecticut: 1993); world printing and writing paper consumption from FAO, AGROSTAT-PC 1993, Forest Product electronic data series, Rome, 1993; doubling of U.S. office paper consumption in eighties from "Leafing Through Europe," *The Economist*, August 25, 1990.

55. Share of paper from recycled fibers is Worldwatch Institute estimate based on wastepaper consumption from Pulp and Paper International, op. cit. note 52, and on world paper production from FAO, op. cit. note 46.

56. Duncan Poore, *No Timber Without Trees: Sustainability in the Tropical Forest* (London: Earthscan Publications, 1989).

57. Stan Sesser, "Logging the Rain Forest," *New Yorker*, May 27, 1991.

58. Ibid.

59. Indonesia from Charles Barber, Nels Johnson, and Emmy Hafild, *Breaking the Logjam: Obstacles to Forest Policy Reform in Indonesia and the United States* (Washington, D.C.: World Resources Institute, 1993); Philippines from Daniel Stiles, "Power and Patronage in the Philippines," *Cultural Survival Quarterly*, Summer 1991.

60. Commission of Inquiry into Aspects of the Timber Industry in Papua New Guinea, *Interim Report No. 4, Vol. 1* (Port Moresby, Papua New Guinea: Government of Papua New Guinea, 1989).

61. Indonesia from Mallet, op. cit. note 41; Philippines from Johnson and Cabarle, op. cit. note 42.

62. Congressional contributions from Barber, Johnson, and Hafild, op. cit. note 59; votes to increase harvest from Andy Stahl, forester, Sierra Club Legal Defense Fund, Seattle, Washington, private communication, October 5, 1993.

63. Dietrich, op. cit. note 27.

64. SILDAP-Sidlakan, press release, Butuan City, Mindanao, Philippines, July 20, 1993.

65. Ansulang, op. cit. note 9.

66. "2000 Species of Fauna Protected" (advertisement), *Washington Post*, April 1, 1993; Mike Harcourt, premier, Province of British Columbia, speech to annual convention of the International Woodworkers' Union, Vancouver, B.C., October 26, 1992.

67. Job creation from value-added manufacturing from Catherine Mater, vice president, Mater Engineering, Ltd., presentation at First North American Conference on Trade in Sustainable Forest Products, Washington, D.C., May 26, 1993.

68. Bill Dietrich, "It's Time To Solve this Mess," *Seattle Times*, April 2, 1993.

_____ 100. **Beyond the Petroleum Age: Designing a Solar Economy** by Christopher Flavin
and Nicholas Lenssen.
_____ 101. **Discarding the Throwaway Society** by John E. Young.
_____ 102. **Women's Reproductive Health: The Silent Emergency** by Jodi L. Jacobson.
_____ 103. **Taking Stock: Animal Farming and the Environment** by Alan B. Durning and
Holly B. Brough.
_____ 104. **Jobs in a Sustainable Economy** by Michael Renner.
_____ 105. **Shaping Cities: The Environmental and Human Dimensions** by Marcia D. Lowe.
_____ 106. **Nuclear Waste: The Problem That Won't Go Away** by Nicholas Lenssen.
_____ 107. **After the Earth Summit: The Future of Environmental Governance**
by Hilary F. French.
_____ 108. **Life Support: Conserving Biological Diversity** by John C. Ryan.
_____ 109. **Mining the Earth** by John E. Young.
_____ 110. **Gender Bias: Roadblock to Sustainable Development** by Jodi L. Jacobson.
_____ 111. **Empowering Development: The New Energy Equation** by Nicholas Lenssen.
_____ 112. **Guardians of the Land: Indigenous Peoples and the Health of the Earth**
by Alan Thein Durning.
_____ 113. **Costly Tradeoffs: Reconciling Trade and the Environment** by Hilary F. French.
_____ 114. **Critical Juncture: The Future of Peacekeeping** by Michael Renner.
_____ 115. **Global Network: Computers in a Sustainable Society** by John E. Young.
_____ 116. **Abandoned Seas: Reversing the Decline of the Oceans** by Peter Weber.
_____ 117. **Saving the Forests: What Will it Take?** by Alan Thein Durning.

_____ **Total Copies**

☐ **Single Copy: $5.00**
☐ **Bulk Copies (any combination of titles)**
    ☐ 2–5: $4.00 ea.   ☐ 6–20: $3.00 ea.   ☐ 21 or more: $2.00 ea.

☐ **Membership in the Worldwatch Library: $25.00 (international airmail $40.00)**
The paperback edition of our 250-page "annual physical of the planet,"
State of the World 1993, plus all Worldwatch Papers released during
the calendar year.

☐ **Subscription to *World Watch* Magazine: $15.00 (international airmail $30.00)**
Stay abreast of global environmental trends and issues with our award-
winning, eminently readable bimonthly magazine.

No postage required on prepaid orders. Minimum $3 postage and handling
charge on unpaid orders.

Make check payable to Worldwatch Institute
1776 Massachusetts Avenue, N.W., Washington, D.C. 20036-1904 USA

Enclosed is my check for U.S. $_____

_____

name   **daytime phone #**

_____

address

_____

city   **state**   **zip/country**

DATE DUE

OCT 05 1994	NOV 16 2009
DEC 05 1994	OCT 27 2010
MAY 13 1995	NOV 29 2010
OCT 12 1995	
MAR 19 1996	
APR 02 1996	
APR 27 1996	
APR 30 1996	
APR 30 1996	
OCT 12 1997	
JUL 7 1998	
OCT 11 1998	
NOV 18 1998	
MAR 04 2000	
APR 22 2001	
DEC 09 2003	
DEC 12 2005	